Wild Life LOL!™
Sloths

This book will turn you upside down!

SCHOLASTIC

Library of Congress Cataloging-in-Publication Data
Names: Herrington, Lisa M., author.
Title: Sloths/by Lisa M. Herrington.
Description: New York: Children's Press, an imprint of Scholastic Inc., 2020. | Series: Wild life lol! | Includes index. | Audience:
Grades 2–3 | Summary: "Book introduces the reader to the world of sloths"—Provided by publisher.
Identifiers: LCCN 2019027505 | ISBN 9780531129814 (library binding) | ISBN 9780531132685 (paperback)
Subjects: LCSH: Sloths–Juvenile literature.
Classification: LCC QL737.E2 H47 2020 | DDC 599.3/13—dc23
LC record available at https://lccn.loc.gov/2019027505

Produced by Spooky Cheetah Press

Book design by Kimberly Shake. Original series design by Anna Tunick Tabachnik.

Contributing Editor and Jokester: Pamela Chanko

Printed in Heshan, China 62

SCHOLASTIC, CHILDREN'S PRESS, WILD LIFE LOL!™, and associated logos are trademarks and/or registered trademarks of
Scholastic Inc.

4 5 6 7 8 9 10 R 29 28 27 26 25 24 23 22 21

Scholastic Inc., 557 Broadway, New York, NY 10012.

Photographs ©: cover, spine: Joel Sartore/National Geographic Photo Ark/National Geographic Creative; cover speech bubbles and
throughout: Astarina/Shutterstock; cover speech bubbles and throughout: pijama61/Getty Images; back cover: Inspired By Maps/
Shutterstock; 1: GlobalP/iStockphoto; 2: Suzi Eszterhas/Minden Pictures; 3 top: Morales/age fotostock; 3 bottom: Suzi Eszterhas/
Minden Pictures; 4: Suzi Eszterhas/Minden Pictures; 5 left: all-silhouettes.com; 5 right: SpicyTruffel/Shutterstock; 6: Suzi Eszterhas/
Minden Pictures; 7: Nacho Such/Shutterstock; 8–9: Kung_Mangkorn/iStockphoto; 10–11: Sean Crane/Minden Pictures; 12–13: David
Tipling/Minden Pictures; 14: Suzi Eszterhas/Minden Pictures; 15: Suzi Eszterhas/Minden Pictures; 16: kjorgen/iStockphoto; 17 eagle:
Gabrielle Therin-Weise/Getty Images; 17 jaguar: Ammit/Dreamstime; 17 snake: Christophe Courteau/Minden Pictures; 18: Ivan Kuzmin/
iStockphoto; 19 top left: Humberto Olarte Cupas/Alamy Images; 19 top right: Humberto Olarte Cupas/Alamy Images; 19 bottom left:
Wollertz/Dreamstime; 19 bottom right: Alain Kubacsi/Biosphoto; 20–21: Juan Carlos Vindas/Getty Images; 22: Suzi Eszterhas/Minden
Pictures; 23 left: Mark Kostich/iStockphoto; 23 right: Suzi Eszterhas/Biosphoto; 24–25: DEA Picture Library/age fotostock; 25 right:
Morales/age fotostock; 26 left: DEA Picture Library/age fotostock; 26 right: Heiner Heine/imageBROKER/age fotostock; 27 left: Suzi
Eszterhas/Minden Pictures; 27 right: Suzi Eszterhas/Minden Pictures; 28 left: Rosa Jay/Shutterstock; 28–29 top: Pascale Gueret/
Shutterstock; 28–29 bottom: Joel Sartore/National Geographic Photo Ark/National Geographic Creative; 29 top right: Mark Payne-
Gill/Minden Pictures; 29 bottom right: Suzi Eszterhas/Minden Pictures; 30 map: Jim McMahon/Mapman ®; 30 inset: Suzi Eszterhas/
Minden Pictures; 31 top: Morales/age fotostock; 31 bottom: Suzi Eszterhas/Minden Pictures; 32: Roaming our way/Shutterstock.

TABLE OF CONTENTS

Can you HANDLE these moves?

MEET THE
SLOW-MOVING SLOTH

Are you ready to be amazed and amused? Keep reading! This book will take you on an upside-down adventure!

LOL!
What did the sloth say to the tree branch? **I'm hooked on you!**

Let's hang out!

At a Glance

Where do they live? → Sloths live in dense, rainy forests in Central and South America.

What do they do? → Sloths spend their lives hanging from tree branches—often upside down.

What do they eat? → Sloths mostly munch on leaves, but also eat twigs, flowers, and fruits.

What do they look like? → Sloths have brown-gray fur, long limbs, and sharp claws.

How big are they? →

HINT: You're bigger. Check this out!

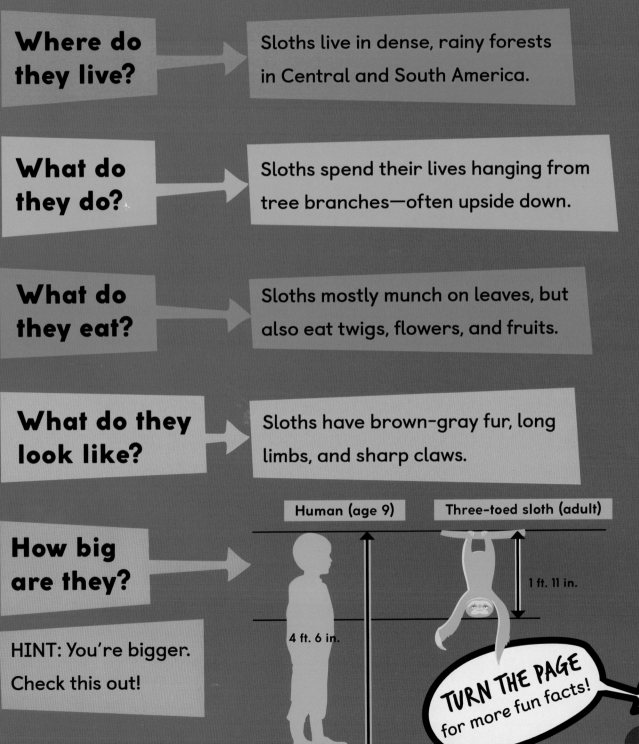

Human (age 9)

Three-toed sloth (adult)

1 ft. 11 in.

4 ft. 6 in.

TURN THE PAGE for more fun facts!

TOE-TALLY DIFFERENT

There are two families of sloths: two-toed and three-toed. Here's how to tell them apart.

What's up?

two claws on front feet

about the size of a medium dog; slightly bigger than three-toed

two-toed

mostly active at night

WACKY FACT: Sloths have really long tongues. They can stick out as far as 12 inches!

LOL!
What does a sloth like to do for exercise? **SLO-ga!**

FAST FACT:
All sloths have three claws on their back limbs. They get their name from the number of claws on their front limbs.

Not much. I'm just hangin' around.

three claws on front feet

mouth shaped like a smile

three-toed

more active during the day

A SLOTH'S BODY

Two- and three-toed sloths have these things in common.

Handy Hooks

Sloths' curved claws are 3 to 4 inches long. The animals use them to hang on to tree branches.

I have a good grip on things!

Is That FUR Real?

A sloth has an upside-down lifestyle. So its thick fur grows opposite that of most **mammals**—away from its stomach. Rain runs right off it!

Long Limbs

Along with their claws, sloths use their four long limbs to move slowly through the trees.

LOL!
What is a sloth's favorite motto?
DON'T HURRY, be happy!

Breathe Easy

A sloth's liver, stomach, and other organs are attached to its rib cage. That keeps the organs from pressing on the lungs—and helps the sloth breathe upside down.

mammals: animals that produce milk to feed their young

LIFE IN THE TREES

Sloths are perfectly suited for a lazy life in **rain forest** trees.

Not SO Lazy

Wild sloths actually snooze only about 10 hours a day. That's not much more than people. A sleeping sloth often tucks its head into its body.

EYE See You!

Three-toed sloths have extra bones in their necks. They can turn their heads almost all the way around.

LOL!
Where do sloths go to school? Elemen-TREE school!

WACKY FACT: Some sloths spend their entire lives in one tree!

rain forest: dense tropical forest where a lot of rain falls

Napping is TREE-mendous!

THAT'S EXTREME! Sloths living under human care may sleep 18 hours a day.

Going Solo

Sloths like living alone most of the time. They just pair up to have babies. And, of course, babies live with their moms for a while!

HOW SLOW ARE SLOTHS?

The sloth is the world's slowest mammal.

Taking Their Time

Sloths travel through the trees at an average rate of about 120 feet a day. It would take a sloth about three days to cover the length of a football field!

The Lowdown

Compared to other animals, sloths have a low body temperature. That helps them save energy.

What's the rush? I'm enjoying the view!

LOL!
What do you get when you cross a sloth with a porcupine? **A slow poke!**

THAT'S EXTREME!
A sloth moving at top speed would be able to travel 70 miles in 20 days. (If it could keep up that pace the whole time!) A cheetah's top speed is 70 miles per hour.

WACKY FACT: The word "sloth" actually means laziness! Look it up!

DOWN BELOW

Check out how sloths move when they are not in the trees.

THAT'S EXTREME! Sloths move even slower on the ground than they do in the trees.

This may take a while!

crawling

Belly Crawl

A sloth's long claws make walking difficult. On land, a sloth drags itself along the ground on its belly to move.

Bathroom Break

Sloths visit the ground only about once a week—to go to the bathroom!

WACKY FACT: A sloth loses one-third of its body weight when it poops.

My favorite stroke is the crawl . . . obviously.

THAT'S EXTREME! Sloths can hold their breath underwater for 40 minutes!

swimming

Cannonball!

During the rainy season, sloths often swim from one tree to another. They have an easy way to get to the water. They drop straight from the trees!

Strong Swimmers

Surprisingly, sloths are good swimmers. They move three times as fast in water as they do when on land.

STAYING SAFE

Trees are the safest place for sloths. They use **camouflage** and their ability to stay still to hide from **predators**, such as eagles, snakes, and jaguars.

> Maybe if I stay VERY still, they won't notice me.

Can't See Me!
Sloths move so slow that a green plant called algae grows on their fur. That helps them blend in with trees.

Stay Back!
When necessary, sloths can use their sharp claws for defense.

camouflage: a way of blending into one's surroundings

predators: animals that hunt other animals for food

YOU ARE WHAT YOU EAT

Sloths mostly eat plant parts, especially leaves. Two-toed sloths sometimes dine on small insects.

THAT'S EXTREME!
It takes a sloth about a month to **digest** a meal! It takes a human six to eight hours.

WACKY FACT:
Leaves are not very nutritious and don't provide sloths with a lot of energy. That's why these animals move so slowly.

LOL!
What is a sloth's favorite dessert? UPSIDE-DOWN cake!

digest: to break down food so it can be used by the body

cecropia fruit

Here are some favorite sloth snacks.

ants

hibiscus flowers

YOU CAN'T HURRY LOVE

Male and female sloths come together to **mate**. Take a look at how they start a family!

LOL!
What is a sloth's least favorite thing to eat?
FAST food!

Back off, pal!

1

Looking for Love

When a sloth reaches three to five years old, it is ready to mate.

mate: to join together to have babies

2 Call of the Wild

A female screams to let males know she is ready to mate. Then she waits for the male to come to her.

3 Fight to the Finish

Males may fight over females. They swipe at each other with their front claws. The female will mate with the winner.

OH, BABY!

After 6 to 12 months, the female sloth gives birth to one baby.

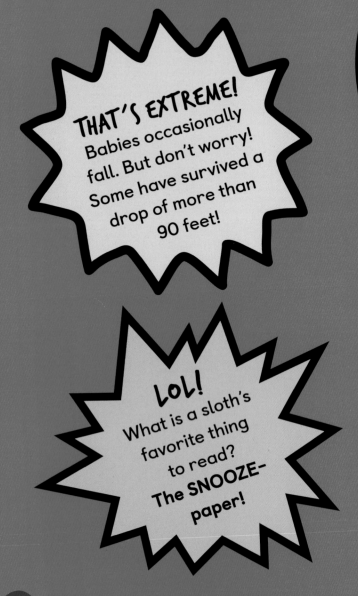

THAT'S EXTREME!
Babies occasionally fall. But don't worry! Some have survived a drop of more than 90 feet!

LOL!
What is a sloth's favorite thing to read? The SNOOZE-paper!

1

Special Delivery

A female gives birth high in the trees. Her newborn is about 10 inches long and weighs about 12 ounces. The baby is born with fur and claws and has its eyes open.

2

3

I'm hungry!

Like all mammals, a baby sloth drinks milk from its mother right after it is born. By the time it is about two months old, the baby is ready to also munch on leaves.

A Clingy Baby

The baby clings to its mom's belly for the first five weeks to nine months. It stays there until it can feed itself and climb on its own. The youngster may live with its mom for six months to more than two years.

ANCIENT GIANTS

Giant ground sloths first appeared about 35 million years ago. Sloths sure have changed a lot since then!

THAT'S EXTREME! Megatherium was the size of an elephant!

Meet *Megatherium*
The largest of the ancient ground sloths was *Megatherium*. It lived in South America.

Gone Forever
Giant ground sloths died out about 11,700 years ago. Scientists think hunters caused the animals to go **extinct**.

extinct: no longer found alive

Big Bones
Megatherium means "giant beast." Its **fossils** were discovered in Argentina in the 1780s.

WACKY FACT: Some ancient sloths even lived in the sea!

That's MEGA-cool!

Um . . . you're mega upside down!

fossils: plants or animals from millions of years ago preserved as rock

SLOTHS AND PEOPLE

We have a long history together.

11,700 years ago

Giant ground sloths roamed throughout the Americas until they died out about 11,700 years ago. They eventually evolved into two-toed and three-toed sloths.

Late 1950s

Some sloths faced dangers from loggers who began cutting down rain forest trees. Loss of habitat and human hunting continue to threaten the survival of certain species.

A caregiver at a sloth rescue center brushes a baby's fur.

THAT'S EXTREME! A sloth's fur can be home to as many as 950 insects like moths!

Today

Some rescue centers help sloths that are injured or have lost their mothers. As soon as the sloths can survive on their own, the workers release the animals back into the wild.

Looking ahead

If logging and other threats continue, more than half of the Amazon rain forest may be gone by 2030. Many groups are working to protect sloths and their homes.

Sloth Cousins

Believe it or not, all anteaters and armadillos are related to sloths!

Anteaters are the sloth's closest relatives.

I'm a type of anteater. But like my sloth cousins, I spend most of my time in the trees.

giant anteater

tamandua

Please note: Animals are not shown to scale.

I'm the only armadillo that can roll into a ball!

three-banded armadillo

I'm the only armadillo that lives in the United States.

nine-banded armadillo

Are you sure we're related?

The Wild Life

Check out this map of the world. The areas in red show where sloths live today: Central and South America. We want the sloths' rain forest **habitats** to hang around for the future. Otherwise, one day there might not be any red left on this map!

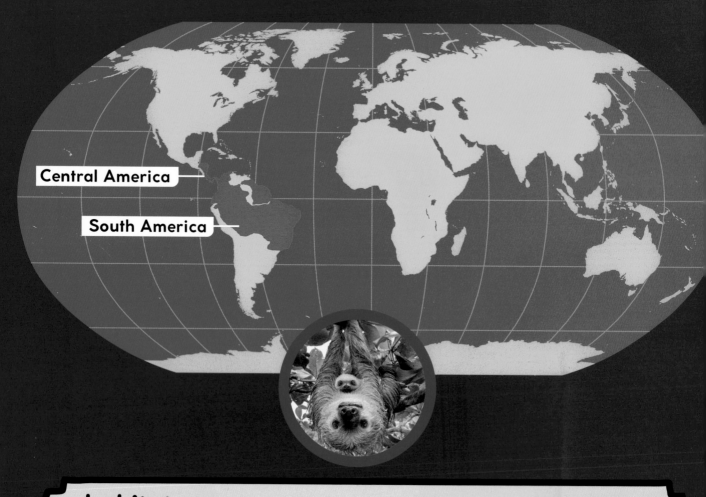

Central America

South America

habitats: the places where a plant or an animal makes its home

What Can You Do?

1 Sloths' rain forest homes disappear as loggers cut down trees for paper, homes, roads, and farms. Each second, an area the size of a football field is cleared from the rain forest. Try to use less paper by writing on both sides and recycling.

2 You know that sloths are great at hiding out. Well, don't hide what you know about them. Speak up to teach others what you learned about sloths so they'll want to protect them, too.

3 To study sloths, some scientists attach tiny backpacks to the animals. The packs have high-tech tracking tools inside. Scientists are using information they gather to find ways to protect sloths. The backpacks detach after a few weeks. Organize a lemonade stand and donate the money to an organization that helps sloths.

ABOUT THE AUTHOR

Lisa M. Herrington has written many children's books about animals—and slowly but surely finished this one. She lives in Trumbull, Connecticut, with her husband and daughter. When she isn't trying to make them laugh, Lisa enjoys reading and traveling.

Did you like the book? Fingers crossed!